A Question of Science

why can't Penguins Fly?

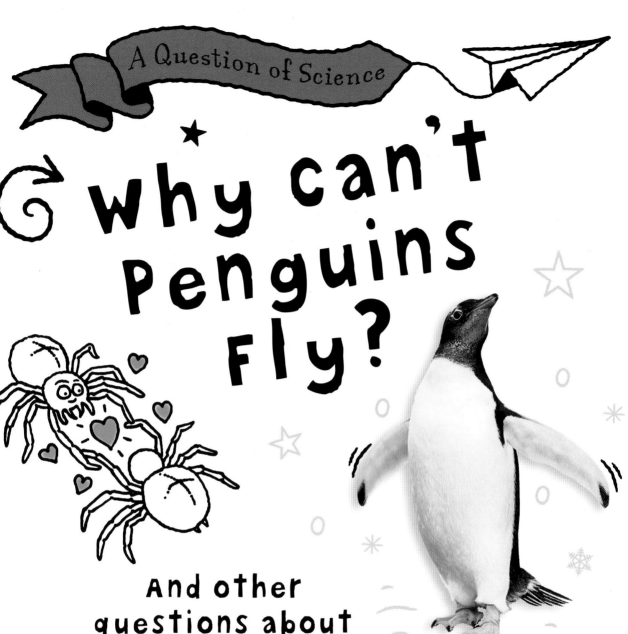

And other questions about ANIMALS

Anna Claybourne

CRABTREE
PUBLISHING COMPANY
WWW.CRABTREEBOOKS.COM

CRABTREE
PUBLISHING COMPANY
WWW.CRABTREEBOOKS.COM

Published in Canada
Crabtree Publishing
616 Welland Ave.
St. Catharines, Ontario
L2M 5V6

Published in the United States
Crabtree Publishing
347 Fifth Avenue
Suite 1402–145
New York, NY 10016

Published in 2021 by Crabtree Publishing Company

First published in 2020 by Wayland
© Hodder and Stoughton 2020

Author: Anna Claybourne

Editorial Director: Kathy Middleton

Editor: Julia Bird

Proofreader: Petrice Custance

Design and illustration: Matt Lilly

Cover design: Matt Lilly

**Production coordinator and
Prepress technician:** Tammy McGarr

Print coordinator: Katherine Berti

Printed in the U.S.A./082020/CG20200601

Picture credits
Alamy: Arco Images 11tc; Everett Collection 17t; Hemis 26t; Christina Kennedy 27bc; Life on white 18b; Minden Pictures 13t, 13br; Science Picture Co 5c; Phil Willis 4cl.
Dreamstime: Yujie Chen 24c; Eric Isselee 22tr; Pius Lee 23tr; Jill Peters 12tr.
iStock: Dissoid 20bcl.
Nature PL: Franco Banfi 9bc.
Science Photo Library: Mint Images-Art Wolfe 22c.
Shutterstock: A B Photographie 28t; Ton Bangkeaw 27br; Be Good 23tl; Audrey Snider-Bell 10t; bernatets photo 4tr; Bhathaway 15c; Blue Crayola 14cl; Mariia Boiko 13cl; Andrew Burgess 7tr; Butterfly Hunter 15tc; Cherdchai Chaivimol 11cr; Chros 14cr; Ethan Daniels 5bl; Chase Dekker 26cl; David Denby Photography 20bl; Dede Dian 10bl; Diyana Dimitrova 9c; DN1988 10br; Feathercollector 7tc; Peter Fodor 5tl;Die Fotosynthese 10bc; Four Oaks 26cr; Giedriius 4b; HollyHarry 13cr; Herschel Hoffmeyer 26br; hsagencia 7bl; Vitalii Hulai 6b;Irin-k 5tcr, 7tl; Eric Isselee 13bc, 20c; Brian A. Jackson 4tc; Jarvaman 1; Petr Jan Juracka 12bl; Tory Kallman 4cr; Kavcicm 9cr;Ibrahim Kavus 14tc; Breck P Kent 15tr; Ivan Kuzmin 20br; Elise Lefran 23b; Leungchopan 5tcl; Joao Luiz Lima 29b; Marques 17cr; Mehmetkrc 21tl; Frances van der Merwe 13bl;John Robert Miller/AP/Rex 17b; Przemyslaw Muszynski 14tl; Nechaevkon 11br; F. Neidl 5cl; Nick 626 14bl; Alta Oosthuizen 9t;Pakhnyushchy 28b; PhotocechCZ 10cl; PHOTO FUN 14clb, 14crb; photok.dk 4tl; Protasov AN 9br; Khongsak putaha 8t; Scott E Read 10c; Revel Pix LLC 29c; RLS Photo 7c; Michael Rosskothen 26bl; Chuck Rousin 8c; RugliG 11tr; Saulty72 10cr; Susan Schmitz 17cl; 7th Son Studio 24b; kowit sitthi 14br; Connor Skye 16b; Steven Russel Smith -Ohio 15tl; Solar seven 27t; Ana Subbotina 29t; Sunnybunny Studio 5tc; Ferdy Timmerman 20bcr; Joost van Uffelen 19t;1Marco Uliana 11crb; Ye Choh Wah 5bc; Richard Whitcombe 16c; wildestanimal 5tlc; Vladimir Wrangel 5tr; Yevgeniy11 19c; Alex Zabusik 9cl; Oleg Znamenskiy 19c.

Library and Achives Canada Cataloguing in Publication

Title: Why can't penguins fly? : and other questions about animals / Anna Claybourne.
Names: Claybourne, Anna, author.
Description: Series statement: A question of science | Includes index.
Identifiers: Canadiana (print) 20200254626 | Canadiana (ebook) 20200254669 | ISBN 9780778779087 (softcover) | ISBN 9780778777083 (hardcover) | ISBN 9781427125408 (HTML)
Subjects: LCSH: Animals—Juvenile literature. | LCSH: Animals—Miscellanea—Juvenile literature. | LCGFT: Trivia and miscellanea.
Classification: LCC QL49 .C53 2020 | DDC j590—dc23

Library of Congress Cataloging-in-Publication Data

Names: Claybourne, Anna, author.
Title: Why can't penguins fly? : and other questions about animals / Anna Claybourne.
Description: New York, NY : Crabtree Publishing Company, 2021. | Series: A question of science | First published in 2020 by Wayland.
Identifiers: LCCN 2020023608 (print) | LCCN 2020023609 (ebook) | ISBN 9780778777083 (hardcover) | ISBN 9780778779087 (paperback) | ISBN 9781427125408 (ebook)
Subjects: LCSH: Animals--Juvenile literature.
Classification: LCC QL49 .C599 2021 (print) | LCC QL49 (ebook) | DDC 590--dc23
LC record available at https://lccn.loc.gov/2020023608
LC ebook record available at https://lccn.loc.gov/2020023609

Contents

What are animals?

Animals are living things. They are important to us in many different ways.

As our pets

SQUEAK, SQUEAK, SQUEAK!

As farm animals, which provide us with things such as food and wool

ME FIRST!

Ladybug

As wildlife we see in parks and gardens

As rare animals in the wild

As animals we visit in zoos

Polar bear

Killer whale

What makes it an animal?

You might think you know an animal when you see one, but what exactly is an animal? Here's a handy checklist:

① Animals get their energy by eating food, whether they eat plants or other animals.

② They breathe by taking in oxygen from the air, or from water, as fish do.

③ They are good at sensing their surroundings and reacting fast, such as this hungry eagle.

Eyes

Nostrils

Wings

Eagles eat fish and other small animals

④ Most animals can move around. Some walk, some run, some fly, and some swim.

4

All kinds of animals!

Animals come in a huge variety of types,
some of them pretty peculiar!

Vertebrates

Animals with backbones
and skeletons.

Elephant

Shark

Human
being

Invertebrates

Animals without skeletons, but
who may have a hard outer shell.

Snail

Fly

Octopus

Invertebrates include some tiny, nearly invisible animals, such as these:

Tardigrade or
water bear

HELLO, I'M DOWN HERE!!

Nematode worm

SMALL IS BEAUTIFUL!

Actual size

... as well as some strange creatures that don't look very much like animals!

Sponge

Feather
duster worm

The **animal kingdom** contains quite
a few surprises and mysteries, too.
Can cockroaches really live without
their heads? Which animal is better
at changing color than a chameleon?
Which animal can you have a chat with?
And just what's inside a camel's hump?

Read on for the answers!

Do spiders have a heart?

YES! Spiders do have hearts, just as most other animals do. But there are a few heartless animals out there.

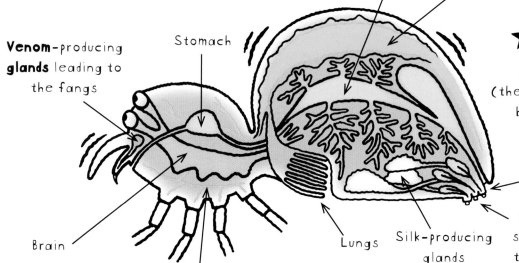

I LOVE HOLDING LEGS WITH YOU!

Inside a spider

If you think spiders are a bit gross, you'll probably find this picture even more disgusting—it's a spider's insides!

Intestines

Heart

Venom-producing **glands** leading to the fangs

Stomach

Anus (the spider's butt!)

Brain

Blood vessels

Lungs

Silk-producing glands

Spinnerets, where spider silk comes out to form a web

What are hearts for?

Hearts pump blood around an animal's body, to deliver useful things such as oxygen and food **chemicals** to its **cells**. Humans are much bigger than spiders, and we have millions of **blood vessels**.

The human heart is slightly to the left side of the body.

A spider's heart is right in the middle of its back.

Blue-blooded

Why is the spider's blood shown here in blue? Because spider blood IS blue. It contains different chemicals than human blood. When a spider's blood is carrying oxygen, it's a pale blue color.

And did you know...

Horseshoe crabs also have blue blood.

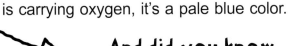

Horseshoe crab

Beetles have yellow blood.

The **ocellated** icefish actually has see-through blood!

Heartless!

Animals that do not have hearts include jellyfish, flatworms, and sponges. Oxygen and **nutrients** just spread through their bodies without being pumped.

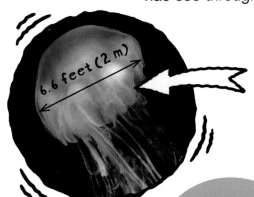

6.6 feet (2 m)

The lion's mane jellyfish is probably the biggest animal with no heart.

Lots of hearts

Then there's the squid who has three hearts, and the earthworm who has five!

Earthworm

Five hearts

Biggest heart

The prize for the biggest heart goes to the biggest animal, the blue whale. Its heart is about the size of a sofa and weighs about 440 pounds (200kg)—as much as three people.

How can a snake swallow a deer?

MMMM, YUM, YUM!

Snakes kill their **prey** with a venomous bite, or by coiling around it and squeezing it to death. Then they have to eat it, but snakes have no limbs. This makes it hard for them to hold prey and tear it apart. Instead, they swallow it whole.

A python swallowing a deer

But how?

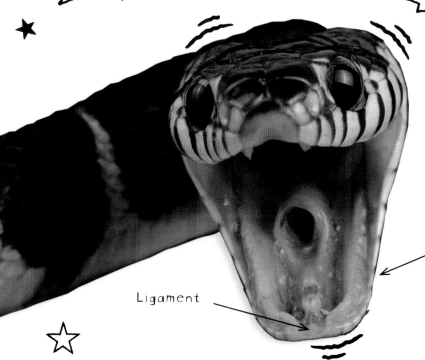

Ligament

That deer is at least twice as wide as the snake's head! To be able to swallow big prey, snakes have **evolved** special jaw bones:

The lower jaw has a gap at the bottom, connected by stretchy **ligaments**. This allows the jaws to open extra wide.

The snake can stretch its mouth and body to wrap around the prey. It moves forward bit by bit until it has swallowed it all.

TIME FOR A SNOOZE... ZZZZ.

Now what?

Afterward, the snake can't move. It has to just lie there for several days while its body **digests** and dissolves the meal.

While it's digesting its prey, the snake risks being attacked by other **predators**. If this happens, snakes regurgitate, or throw up, what they've just eaten, so they can escape!

Are snakes just greedy?

No—snakes eat like this because it can be hard to find food. When they do find food, they gobble up as much as possible, to keep them going until the next kill. Many other predators do this, too, including wolves, big cats, and sharks.

But there are other ways to make sure you have enough food. Store it!

Squirrels dig holes to store nuts.

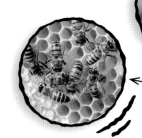

Honeybees store their honey in their nests.

Spiders wrap up prey in silk and save it for later.

CAN'T WAIT!

Live on it!

Parasites are even smarter. They live on or inside another living animal and feed on it.

This copepod lives on a Greenland shark's eyeball, and nibbles it when it feels hungry.

EEEEwwW!

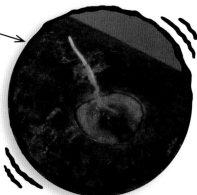

Head lice live in our hair and feed on our blood!

Which animals eat people?

Every year, humans eat billions and billions of animals. We eat them far more than they eat us! However, animals do sometimes eat people, and the animals most likely to eat humans are:

Crocodiles

They are attracted by people who are swimming or paddling.

Big cats

Especially lions, tigers, and leopards

Bears

Polar bears, black bears, and grizzlies can all eat people.

Sharks

They're not as dangerous as people think, but they do eat a few people each year.

Snakes

Many people die from snake bites, but snakes swallowing humans is extremely rare.

Wolves and wild dogs

They hunt in packs and will sometimes eat a human.

Pigs

Yes! pigs have occasionally been known to eat people, usually farmers.

Oops!

It's very rare for animals to actually hunt humans as their main prey. If an animal eats a human, it's usually because it's desperate for food or mistakes the human for something better-tasting.

Seal

Surfer

Great white sharks are a good example. Their favorite food is seals...

...but sometimes they attack surfers, because a surfer paddling a surfboard resembles a seal from below.

Top of the food chain

Animals that eat humans are usually **apex** predators, These are animals at the top of the **food chain** or **food web**.

In an African **savanna** food chain (left), a lion is the apex predator. It eats animals such as deer, which eat plants. It is not eaten by other animals. In **prehistoric** times, humans would have been part of a savanna food chain (right). Today we mostly live in houses and can avoid big predators. Also many predators are **endangered** and rare. That's why very few humans actually do get eaten.

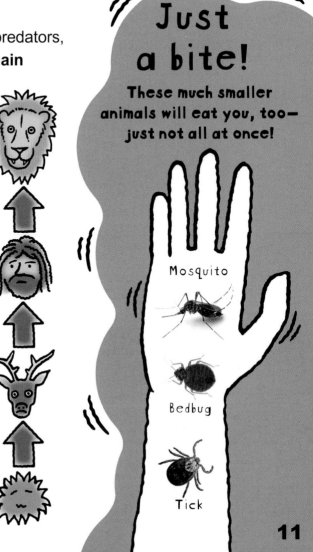

Just a bite!

These much smaller animals will eat you, too— just not all at once!

Mosquito

Bedbug

Tick

Why don't cats lay eggs?

Have you ever seen a cat egg? Probably not, because cats don't lay eggs! Like dogs, elephants, and humans, cats give birth to live babies.

Imagine if cats did lay eggs, though!

Would they be furry?

MEOW!

Kittens would hatch out, like this.

Crab with her eggs

Who lays eggs?

In the animal kingdom, laying eggs is the most common way to reproduce, or have babies. All birds lay eggs. So do most fish, **reptiles**, **amphibians**, **insects**, and sea creatures, such as octopuses and jellyfish.

Why lay eggs?

An egg is a supply of food with a protective shell or covering. The egg lets the baby grow and develop outside the mother's body.

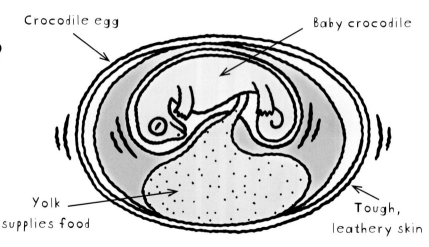

Crocodile egg

Baby crocodile

Yolk supplies food

Tough, leathery skin

Many egg-laying animals lay lots of eggs. This means there's a better chance that some of their babies will survive. Egg-laying also means the mother isn't weighed down by babies growing inside her—a big advantage for flying birds.

A salmon can lay up to 20,000 eggs!

So why NOT lay eggs?

Most **mammals** have live babies instead of laying eggs. There are advantages to reproducing this way. The baby is safer and warmer inside the mother. There may be fewer babies, but mammals look after their young and feed them milk. This care and protection gives babies a better chance of surviving.

SLURP! SLURP! SLURP!

There are also some lizards, snakes, fish, and even insects that don't lay eggs.

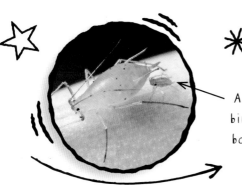

Aphids give birth to tiny baby aphids!

World's weirdest eggs!

Horn shark eggs are shaped like a corkscrew.

A few mammals actually do lay eggs: platypuses and echidnas.

Echidna egg

Echidna

So, if cats did lay eggs, they might look like this!

Why don't caterpillars look like their parents?

A baby human looks like a small version of an adult human. A kitten looks like a small cat. But a caterpillar looks nothing like its parents!

Mini-cat

A red admiral caterpillar...

... becomes this—an adult red admiral butterfly.

Transformers

Many animals change like this through their lives, especially smaller animals. This change is called **metamorphosis**.

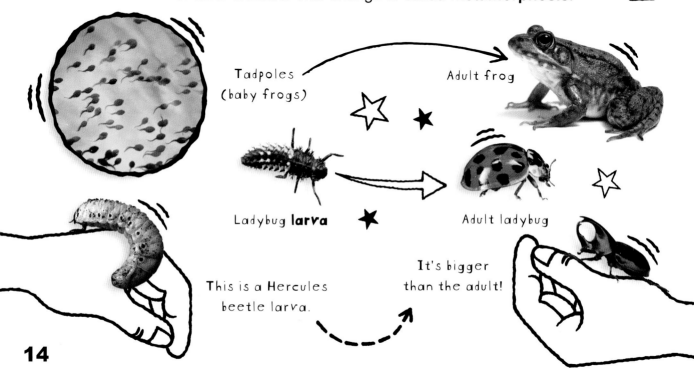

Tadpoles (baby frogs)

Adult frog

Ladybug **larva**

Adult ladybug

This is a Hercules beetle larva.

It's bigger than the adult!

14

Why change?

Metamorphosis usually happens in animals that don't care for their babies after they are born. The babies are left to survive on their own. The babies and adults often live in different places and eat different food, so they need different bodies.

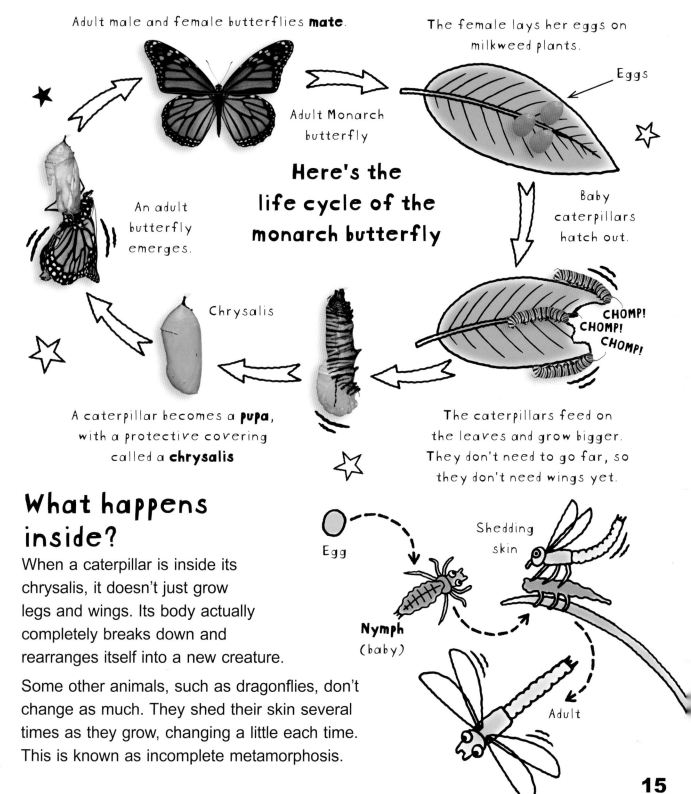

Adult male and female butterflies **mate**.

Adult Monarch butterfly

The female lays her eggs on milkweed plants.

Eggs

Here's the life cycle of the monarch butterfly

Baby caterpillars hatch out.

CHOMP! CHOMP! CHOMP!

An adult butterfly emerges.

Chrysalis

A caterpillar becomes a **pupa**, with a protective covering called a **chrysalis**

The caterpillars feed on the leaves and grow bigger. They don't need to go far, so they don't need wings yet.

What happens inside?

When a caterpillar is inside its chrysalis, it doesn't just grow legs and wings. Its body actually completely breaks down and rearranges itself into a new creature.

Some other animals, such as dragonflies, don't change as much. They shed their skin several times as they grow, changing a little each time. This is known as incomplete metamorphosis.

Egg

Shedding skin

Nymph (baby)

Adult

15

Why can't animals talk to us?

Wouldn't it be amazing if animals could talk as we can, and have conversations with us? You could have a long chat with your pet or with an elephant in the zoo!

But you can't!

Humans are the only animals (as far as we know) who talk using thousands of words and long sentences.

Our **complex** brains can learn a lot of words and language rules.

Our specially shaped throats and tongues can make different word sounds.

Animal language

However, a lot of animals can communicate. They send each other signals and messages using sounds, smells, actions, or colors.

Cuttlefish flash patterns of colors to each other on their skin.

HANDSOME MALE OVER HERE, LOOKING FOR A MATE!

Cuttlefish

Vervet monkeys make different sounds to warn each other about different predators. Their sounds are like simple words.

GRUNT, GRUNT!*

SNORT!**

HONK!***

*Look out, a leopard!

**Look out, a snake!
***Look out, an eagle!

And some animals CAN actually talk to us—kind of!

Sign language

Apes are quite clever. Some, such as chimps, bonobos, and gorillas, can learn sign language. A bonobo named Kanzi has learned over 300 word signs, and can point to them to say things, such as "Cook marshmallows."

> I WANT TO PLAY WITH YOU!

Best friends

Dogs can learn some human words too, such as "walk" or "food." They can also tell us things with their bodies and faces, such as:

> IT WASN'T ME WHO ATE THE SAUSAGE.

Humans have **tamed** and **bred** dogs for thousands of years, which may be why we understand each other so well.

Top talkers

Parrots are probably the best animal talkers of all. They can make speech sounds with their mouths, and often copy what we say. But do they know what they're saying? Some parrots do seem to understand what words mean, and can say things to us.

> YOU BE GOOD, SEE YOU TOMORROW!

A famous African gray parrot, Alex, could say over **100** words and make up short sentences.

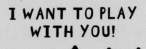

17

Which is the smartest animal?

THAT WAS EASY!

Other than humans, it's hard to say which animal is the smartest because each one is intelligent in different ways. There are several in the running for the "smartest animal" award.

Signs of high intelligence include:

- Recognizing themselves in a mirror
- Remembering things
- Solving problems
- Using tools
- Playing

Humans do all these things. These animals can do some of them:

Chimpanzees

- Chimps invent tools for different situations. For example, they use small sticks to catch termites and big sticks for mashing hard potatoes.

- They beat humans at remembering where numbers appear on a screen.

- A chimp knows it can see itself in a mirror. If it sees a mark on its face it will try to rub it off.

Chimps can recognize themselves in a mirror.

Dolphins

Dolphins live in groups and communicate with squeaks and clicks.

- They love playing. For example, blowing bubble rings for fun.

- They use sponges like a tool to protect their noses from sharp **coral**.

- They're sneaky. An aquarium dolphin named Kelly received a fish for each piece of litter she collected from her tank. So she hid a large piece and tore it into smaller bits to get more fish!

Elephants

They say elephants never forget—and they really do have amazing memories.

- The oldest female elephant memorizes the location of distant watering holes and leads her herd to them during dry spells.

- They can recognize the voices and languages of people who've hunted them.

- They help each other with problems.

Crows

So clever, it's creepy! Crows play catch, recognize different humans, and use tools.

- Crows can use a hook to get food.

- If they don't have a hook, they can make one.

- They can plan ahead and solve problems, as in this experiment:

To get the snack, crows drop stones into the jar to make the water level higher. Genius!

Narrow jar

Floating snack

Water

Stones

How do chameleons change color?

Chameleons are famous for changing color to match their surroundings. We even call a person who changes to suit the situation they're in a chameleon.

But chameleons don't actually do this at all!

YOU'VE GOT IT ALL WRONG!

Chameleons can change color. They do it quite slowly—not in a split second. But they don't do it to match their surroundings.

Chameleon colors

So why do they do it? There are several reasons.

A chameleon will change to stronger or darker colors if it's angry or scared.	When males fight each other, they have vivid patterns.	Chameleons use bright color patterns when they're looking for a mate.	Some chameleons turn darker when they want to get warmer. A dark surface absorbs more heat from the Sun.

Keeping it green

Most chameleons are green when they're relaxed. This gives them some **camouflage** to hide in their jungle home.

When a chameleon changes back from a bright color to green, it could look as if it's doing it to blend in. In fact, it's just calming down!

How does it work?

A chameleon's skin has several layers.

The outer layer has yellow **pigments**.

Underneath is a layer of tiny crystals.

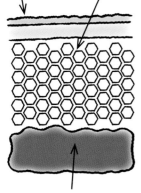

Under that is a layer containing a dark chemical called **melanin**.

Normally, the crystals reflect blue light. It mixes with the yellow layer to make green.

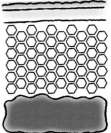

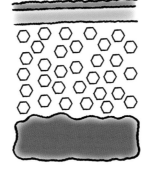

To change color, the chameleon moves the crystals farther apart. This makes them reflect green, yellow, orange, or red. These combine with the yellow layer to make a range of colors.

To become darker, the chameleon releases melanin from the lowest layer into the upper layers.

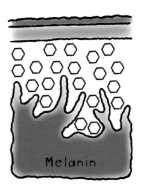

Melanin

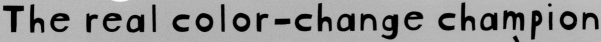

The real color-change champion

However there IS an animal that changes color in a flash to match its surroundings. It's the octopus! Octopuses have dots of pigment in their skin. They can quickly shrink or stretch the dots to make different colors and patterns.

What's inside a camel's hump?

Camels live in dry desert areas, where there isn't much water. So, some people think camels store water in their humps.

A camel can drink a LOT of water at once—up to 53 gallons (200 l). That's about 800 normal-sized glasses of water. But it doesn't get stored in the camel's humps!

So what's in there!??

Desert-dwellers

As well as being short of water, desert areas are also short of food. Camels can walk for days without drinking OR eating. How do they do it?

Most camels are pack, or work, animals, used for riding or carrying heavy loads.

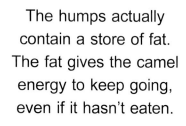

Hump

The humps actually contain a store of fat. The fat gives the camel energy to keep going, even if it hasn't eaten.

Fat

Energy

Droopy humps!

When a camel's hump has a store of fat inside, it sticks up. When the camel has used up the fat, the hump droops down.

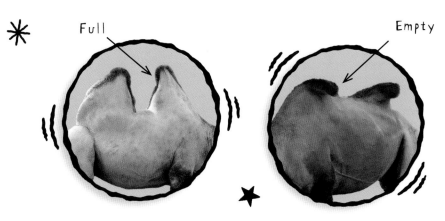

Full

Empty

Where does the water go?

Camels are quite big, so when they fill up on water, they spread it throughout their bodies, carrying it in their blood, organs, and skin. Then they gradually use it up.

One hump or two?

There are two types of camels: Bactrian and dromedary. Bactrian camels have two humps, and dromedaries have one. The humps work the same way for both.

To remember which is which, just look at the first letter:

Bactrian = two humps!
Dromedary = one hump!

No humps!

Baby camels don't have much of a hump at all. They only grow their humps when they start eating solid food.

Desert survivor

Camels have other ways of surviving in the desert, too:

- They can close their nostrils to keep sand out
- They save water by not sweating much
- They have very dry poop that doesn't contain much water.

Camel poop

How can a cockroach live without its head?

It's a famous fact that cockroaches can live without their heads.

How is that possible!?

You definitely need your head. It contains your brain, which is in charge of your body. You need your mouth and nose for eating and breathing. When people lose their heads, it's bad news.

That's why, long ago, people were beheaded as a way of executing them.

OFF WITH HIS HEAD!

But cockroaches are a bit different.

Head

Thorax

Abdomen

Cockroaches are insects, and have six legs.

Meet the cockroach!

There are thousands of cockroach **species**, and only a few of them invade human homes. The German cockroach is one of them.

Life without a head

If, for example, a bird bites off a cockroach's head, it really CAN survive—for a while. That's because a cockroach's body works differently than ours in several ways:

MMM, CRUNCHY!

① Cockroach blood isn't under high **pressure** like ours. If the head is pulled off, the neck doesn't leak blood. It just heals up.

Blood seals at the neck.

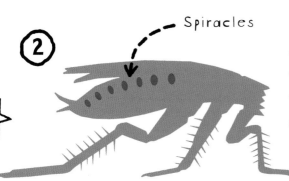

Spiracles

② Cockroaches don't breathe through their mouths. They take in air through holes on their bodies, called spiracles. So they can still breathe!

③ A cockroach has a brain in its head, but it also has smaller mini-brains, called ganglia, along its body. The ganglia control its legs and organs, so it can live without its main brain.

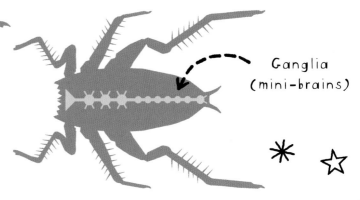

Ganglia (mini-brains)

Dead cockroach

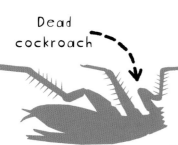

④ Cockroaches do need to eat and drink, and that is their downfall. They can go without food for a month or so, but they can only manage without water for about a week. So that's how long a headless cockroach can live.

Creepy cockroach facts!

- A cockroach can hold its breath underwater for 30 minutes.
- Cockroaches give off a stinky smell.
- To run fast, cockroaches stand up on their back legs!

Why can't penguins fly?

Penguins are experts at swimming, waddling, and sliding around on the ice— but NOT flying.

They can't fly at all and here's why!

WHEEEE!

Mini wings

Penguins do have wings, but they are quite small and stubby compared to their large, plump bodies.

So why have wings at all?

Long ago, birds evolved wings for flying. Flying was very useful because it helped them find food and escape from danger. Over many **generations**, penguins and other types of birds lost the ability to fly. They still have their wings, but they use them differently.

Gentoo penguin

Small, short wings

This red-throated loon is closely related to penguins, but it can fly.

Its wings are much bigger and longer compared to its body.

How did this happen?

Penguins need to swim fast, steer, and dive underwater to catch food and escape from predators. When penguins swim underwater, they do look like they're flying! They use their wings in a similar way to birds in flight.

Over time, penguins' wings became smaller and more flipper-like to help them swim better.

Penguins are also quite heavy, so these smaller wings can't get them off the ground!

Emperor penguins

Thick blubber for keeping warm

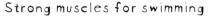

Strong muscles for swimming

We can't fly either!

There are several other types of flightless bird.

Big, fast-running birds such as ostriches and emus

Kiwis are chicken-sized birds that have almost completely lost their wings

The kakapo is the world's only flightless parrot

Quick-fire questions

How do squirrels remember where they hid their nuts?

Squirrels are quite intelligent and have good memories. They have evolved to store food and keep a "map" in their minds of where they put it. However, they do forget about some of their stashes. The forgotten nuts and seeds they have buried often grow into new trees, which is also good news for squirrels!

Why do whales sing?

Some whales sing long, complicated "songs" to attract a mate. Sound travels very well through water, so other whales far away can hear them. The top singers are humpback whales, who learn their songs from each other and make up new ones.

Do animals laugh and cry?

Scientists think humans are the only animals that cry when they feel sad. Other animals do have tears, but they are just for washing out their eyes and keeping them moist. However, a few animals do seem to laugh, including chimpanzees and other apes. And rats make a giggling sound when they are tickled!

TICKLE
TICKLE

STOP IT!

Why don't insects have bones?

Insects, and other animals without bones, evolved before animals with skeletons did. Skeletons are useful for supporting larger land animals, but many smaller animals and sea creatures don't need them. Instead, they have a soft body and a shell, or a strong outer skin called an **exoskeleton**.

Can owls really spin their heads around?

Most owls can turn their heads right around to look straight behind them. They can also turn their heads upside down! This is because owls actually have quite long, flexible necks under their thick feathers. But they can't spin them all the way around. The most an owl can turn its head is about three-quarters of the way around.

WHAT ARE YOU LOOKING AT?

Why do cats lick their butts?

For cats, licking themselves is the best way to get clean. When a cat licks its butt, it will probably swallow some germs. The cat's saliva and stomach contain germ-killing chemicals, so this isn't usually harmful.

Glossary

animal kingdom A category of nature that includes all living things except plants

amphibians Animals that can live both on land and in water

blood vessels The tubes through which blood flows throughout the body

bred The planned mating of certain animals in order to produce specific characteristics in the offspring

camouflage The ability of animals to blend in with their surroundings and not be seen

cells Basic building blocks of life

chemical A substance that cannot be broken down into different parts

chrysalis A type of pupa with a hard outer case

complex Not easy to learn or understand

coral A soft-bodied animal that lives in a stony skeleton in the ocean

digest When the body breaks down food in order to absorb nutrients

endangered A living thing that is in danger of dying out and becoming extinct

evolved Changed over time, as in a living thing

exoskeleton A hard outer shell or skin found in many invertebrates, such as beetles

food chain A pattern of eating and being eaten, beginning with plants and ending with large animals

food web A group of connected food chains within a community

generations Periods of time when groups of people were born and lived

glands A group of cells or an organ in the body that produces a substance for the body to use

insects Animals with six legs and a body divided into three parts

larva A baby insect that has a different form from its parents, such as a caterpillar

ligaments Bands of stretchy material connecting bones or parts of a joint

mammals Types of warm-blooded animals that feed their babies on milk

mate To join together to make babies

melanin A dark brown or black pigment that is found in some animals' skin, eyes, fur, and hair

metamorphosis The way some animals change from one form to another during their life cycle

nutrients Chemicals that provide food or essential minerals for a living thing

nymph A baby insect that looks similar to its parents, such as a dragonfly

ocellated A living thing that is covered in eye-like spots

parasites Living things that live in or on another living thing and use it for food

savanna Areas of land that are covered in grass but with only a few trees or shrubs

pigments Naturally occurring color chemicals, such as those found in plants and animals

predator An animal that hunts, kills, and eats other animals for food

prehistoric A time before information was written down

pressure A force pushing something, such as blood, making it move

prey An animal that is hunted, killed, and eaten by other animals for food

pupa A stage in the life cycle of some insects, where they change from a larva to an adult

reptiles Animals that have blood, scaly skin, and lay eggs

species A group of animals that share the same characteristics and can reproduce together

tamed To make something, typically an animal, gentle and obedient

venom A poisonous substance created by some animals, such as snakes

Learning More

Books

Dakers, Diane. *Dian Fossey: Animal Rights Activist and Protector of Mountain Gorillas.* Crabtree Publishing, 2016.

Newman, Aline Alexander and Gary Weitzman. *How to Speak Dog: A Guide to Decoding Dog Language.* National Geographic Children's Books, 2013.

Rodger, Ellen. *Animals on the Outskirts.* Crabtree Publishing, 2020.

Wilsdon, Christina. *Ultimate Predatorpedia: The Most Complete Predator Reference Ever.* National Geographic Children's Books, 2018.

Websites

https://kids.sandiegozoo.org/
Zoo site with loads of animal facts, games, activities, and videos.

https://a-z-animals.com/
Handy information about thousands of different animals.

www.worldwildlife.org/
Wildlife conservation charity with lots of information about endangered species.

https://kids.nationalgeographic.com/videos/amazing-animals/
Fun animal facts, photos, and videos from National Geographic.

www.zooborns.com/
Meet baby animals from zoos around the world.

Index